The Truth About Jelqing and What You Need to Know

Copyright 2015 by Chris Campbell - All rights reserved.

This document is geared towards providing exact and reliable information in regards to the topic and issue covered. The publication is sold with the idea that the publisher is not required to render accounting, officially permitted or otherwise, qualified services. If advice is necessary, legal or professional, a practiced individual in the profession should be ordered.

In no way is it legal to reproduce, duplicate, or transmit any part of this document in either electronic means or in printed format. Recording of this publication is strictly prohibited and any storage of this document is not allowed unless with written permission from the publisher. All rights reserved.

The information provided herein is stated to be truthful and consistent, in that any liability, in terms of inattention or otherwise, by any usage or abuse of any policies, processes, or directions contained within is the solitary and utter responsibility of the recipient reader. Under no circumstances will any legal responsibility or blame be held against the publisher for any reparation, damages, or monetary loss due to the information herein, either directly or indirectly.

The information herein is offered for informational purposes solely and is universal as so. The presentation of the information is without contract or any type of guarantee assurance.

The trademarks used are without any consent, and the publication of the trademark is without permission or backing by the trademark owner. All trademarks and brands within this book are for clarifying purposes only and are owned by the owners themselves, not affiliated with this document.

Table of Contents

Introduction

Chapter 1: Jelqing Defined Through Male Anatomy

Chapter 2: History of Jelqing and Penis Enhancement

Chapter 3: Best Practices in Jelqing

Chapter 4: Advantages of Jelqing

Chapter 5: Disadvantages of Jelqing

Chapter 6: Male Enhancement in the Modern World

Conclusion

Introduction

The pages in this book were developed through extensive research of unbiased information regarding jelqing. In the hotly debated topic of male enhancement, there are many opinions people hold about what is morally correct. However, in this book, I am going to stay away from telling you what you should do, but, rather, what important information you need to know before you start your decision-making process. This book contains proven steps and strategies on how to enlarge the penis using a technique that does not require the consumption of pills, or undergoing expensive surgical procedures.

Males of all species have given a high degree of importance to the size of their penises. In some male circles, the penis can be a sign of superiority, so much so that status can be affected depending on whether a man's penis is considered large or small. The size of a man's penis has also been regarded as a primary contributing factor to a man's sexual prowess and his ability to provide maximum satisfaction to his partner.

Because of the constant focus on the size of the penis, males in different eras have been reported to resort to various methods of improving their sizes. These male enhancement methods have sometimes bordered on torture, which many males would endure for the future pride that would come with owning a huge organ. The men of the 21st century have had better luck than their predecessors in terms of male enhancement, because they can now choose from a wide variety of methods that are neither painful nor dangerous.

I can guarantee that you will find this book useful if you make sure to implement what you learn from the following pages. I also recommend that you take notes while you are reading the book; this will ensure that you get the most out of the information in here, because you will be able to look over the notes of this book even after you've finished reading it. The notes will help you to pinpoint exactly what you need to know, and by writing things down, you will be able to recall specifics and how to handle certain situations when they arise. I want you to feel that you made a purchase that is worth your money.

Lastly, remember that everything in this book has been compiled through research, my own experiences, and the experiences of others, so feel free to question what you have read in this book. I encourage you to do your own research on the things that you want to look deeper into. I created this book due to the high demand of people looking for information on this topic. There is so much misinformation on the internet that it can be hard to find unbiased facts. Most articles contain some type of agenda in regards to what decision you should make, often with financial incentives involved.

I hope you find this book informative!

Chapter 1:

Jelqing Defined Through Male Anatomy

Back in the day, it was said that nothing could be done to improve the length or girth of one's penis—but not anymore.

In fact, that kind of assumption is only made by people who do not believe they can still improve their masculinity. As you may know by now, the size of a man's penis is important to him, and that's why it is important to make use of exercises that can potentially improve not only a penis' size, but also a man's confidence and belief in himself.

This is where Jelqing comes in...

Stress and Improvements

Jelqing is defined as the process in which a man provides stress to the muscles and tissues of his penis by means of certain exercises, so that in turn, the tissues would stretch and be able to give the man the kind of length and girth he desires (or at least more than he had at the start).

Recent developments in muscle enlargement techniques have shifted the focus from the abdominals, arms, and legs to the muscles of the human penis. Various types of penile enlargement techniques have been developed and tested by males throughout the world.

One of the most well-known methods for penile enlargement is referred to as "jelqing therapy" or just "jelqing." Jelqing is defined by specialists as a method of enlarging the penis through a series of physical exercises that can be performed either at home, or in a male enhancement clinic. These exercises serve to enhance the circulation and pressure of the blood flow to the penis, thus enlarging it.

The largest a penis can grow through jelqing depends on the following factors:

The amount of repetitions performed through the exercise. The more you massage your penis, the more it is forced to adjust to that kind of mechanism; therefore, it is forced to adapt to the stimuli.

Any variations incorporated into the usual technique, such as the use of external jelqing devices. These include penis pumps, caps, and the like.

The level of the male's tolerance while performing the jelqing exercises. And of course, a man's tolerance is definitely part of the equation. It's all about being committed to what you're doing and knowing how much stress you could reasonably provide to your penis each day. More importantly, it's about how much pain or stress you can handle, because the truth is, not everyone has high tolerance for stress, especially when it's being placed on a sensitive area like the penis.

The Importance of Tolerance

Tolerance is a very important factor in regard to jelqing, because jelqing involves a method of therapy quite similar to male masturbation. However, jelqing is differentiated from male masturbation in that ejaculation is not the immediate and primary goal of the exercise. Some experts also state the strokes involved in jelqing therapy are more similar to the strokes needed for milking a cow's breasts. This is the reason jelqing is sometimes referred to as "milking" by some of its practitioners.

Jelqing, as a form of male enhancement therapy, is often credited to ancient Arabic science, though evidence of this is still being further researched by modern scientists. The word "jelq", however, has Arabic roots and is translated into the English language as "milk", hence the idea that "jelqing" the penis is tantamount to "milking" it.

Though recent variations have been added by male enhancement therapists, males can still perform the most basic form of the jelqing exercise at home without having to pay a therapist. The steps involve the following:

Make sure to start from the base of the penis going upwards. This is to ensure you get to massage your penis and that the force won't be too overwhelming, so you'll be able to adjust to it.

Stroke from the base, using a squeezing motion. Let it feel natural, but also make it seem like you're pumping at the same time. This will provide the stress the penis needs to feel for it to be enlarged.

Continue stroking until the corona is reached. You have to make sure the intensity of your strokes will increase progressively. The more tension (not pain) you feel, the better.

Start again from the bottom and continue performing these strokes in as many repetitions as possible.

For those who may not be aware, the corona is the part of the penis that surrounds the tip. It is the strip of skin located between the opening in the tip and the so-called "neck" of the penis. In uncircumcised males, the corona covers the slit-like opening on the tip of the penis and acts as a lubricant by keeping the glans wet with mucus. After circumcision, the tip of the penis becomes dry and permanently exposed.

The results of jelqing therapy may also vary, because the part of the penis, known as the glans (the tip), also has varied proportions in different males. In some men, the glans have a much larger circumference than the shaft. It widens even more when the penis is fully erect, such as in preparation for sexual intercourse. This type of penis is often referred to as the "doorknob" or mushroom-shaped penis. In other males, the glans is a bit narrower, which gives the entire penis a probe-like appearance.

Many supporters of the jelqing method believe the penis is enlarged through this method, because the cavities in the penis become filled with blood during the stroking. This, in turn, serves to make the penis larger when erect.

Although many men have come out to stake their support for jelqing therapy as a male enhancement method, the therapy is not backed by any scientific evidence. Additionally, the claim that the blood-filled cavities of the penis make it larger when erect is also not backed by any biological explanation. However, because it is relatively safer than other techniques and because it is much older than any other male enhancement method (as far we know), jelqing has managed to maintain its place at the top of the ladder in male enhancement therapy.

Chapter 2:

History of Jelqing and Penis Enhancement

The history of mankind is as diverse as it is long, and throughout all of recorded history, one theme has transcended all eras: man's fascination with the penis, its size, and length (or lack thereof). Famous playwrights have written hundreds of plays that revolve around the theme of "manhood", although some plays may not show the theme as blatantly as others.

For this reason, it is believed that men have often sought various methods of enhancing the girth of their penises in whatever way possible. There has even been evidence of penile enlargement methods dating as far back as the prehistoric age. Who would've guessed that even the cavemen seemed to have shared the same belief that a bigger penis is much better than the opposite? There is also evidence of men with genital hernia being revered in some parts of Latin America, because they were perceived to have been "gifted" with large penises.

Looking back in time

As mentioned earlier, men have been fascinated with various methods of penis enlargement, no matter where in the world they're from. Let's take a look at the examples below and see what our fellowmen from the olden days did to enlarge their penises:

Paleolithic Days

Over the years, archaeologists have been doing their best to understand the kind of life that Paleolithic men—or in layman's terms, cavemen, used to do with their time. And guess what they found?

Even Paleolithic men wanted their penises to be bigger. They believed that bigger is better, especially when they were men who had titles or were considered "royal", even in their own tribe. Having a larger penis meant he could be considered a good leader, and that's why the illustrations and hieroglyphics found on cave walls denoted that these men had big members.

Here's another interesting discovery: most prehistoric art pieces are actually statues that involve huge penises, as well as sculptures of phalluses.

The Asians

Penis enlargement also has Asian, particularly Chinese, roots. As you may know, the ancient Chinese were reliant on herbs and had a cure for just about anything. According to historians, the Chinese made a mix of herbs that they once used for penis enlargement. These herbs include the following:

Cuscutta

With its nitric oxide production, mixed with intracavernosal pressure, men reportedly got a fuller and harder erection, in addition to having improved penis size.

Cistanche

Meanwhile, Cistanche is responsible for the improvement of blood circulation, since it contains a lot of alkaloids, meaning it has great anti-inflammatory benefits.

Ginseng

Ginseng enhances blood circulation with the help of nitric oxide.

Fo Ti

According to the discoveries, Fo Ti reduces the risks of penile premature aging and helps promote better hormone and tissue production.

Deer Antler

This helps repair damaged nerve endings in the penis and could also improve the levels of DHEA, Testosterone, and HGH, which are all significant in lovemaking.

They used to rub a mixture of these herbs on their penis before proceeding to massage or put some tension on it.

And here's another interesting—but odd—fact:

Archaeologists and historians agree that some of these men, who didn't believe in the power of herbs, turned to eating the testicles and penises of animals they considered strong, hoping they'd be able to gain the strength of the said animals. Rumors of surgeries by quack doctors abound, too.

It just goes to show that, even in the olden days, many men were willing to do everything and anything to have the penis size they wanted!

The Arabians

Here's the thing: Jelqing is said to have real Arabian roots. In fact, it is said, even from childhood, Arabians were taught that having a big manhood equates to having a great life, and one cannot be considered a man unless his penis grows big.

One of those things that children were taught is to treat the penis like they are milking a cow. This means, after being intense and seeing the erection grow, they must cease action in order for the penis to be flaccid before working on it again. It really does take a lot of patience, but it's said this ensures jelqing works.

Tribes and the Like

There are other interesting—and wildly horrific—acts that would make you realize how far people would go just to improve the state of their penis. For example, Ancient Borneans were believed to have maintained permanent erections by injecting metal rods into their penises! This is said to be the predecessor of the way men would pierce or put certain accessories on their penises, just for it to be enlarged.

Then, there is also this tribe in Brazil, called Topinama, who, back in the 19th century, tended to manipulate snakes, so it would bite them and engorge their penises—which, come to think about it, is quite a dangerous and risky act. And it really was: the men endured pain for 6 whole months, just so they could feel like their penises were bigger than before!

Time for Industrialization

Of course, as times progressed, penile enlargement methods progressed as well. Early mechanical jelqing devices were introduced to the public. However, these devices were not loved by the general public and really did not last long, because they were quite uncomfortable to use.

And then came Otto Ledever. Often credited as the Father of Penis Enlargement, Ledever always had an active interest in sex and things that could make it better. With a background in engineering and knowledge in vacuum technology, he was able to create the first-ever penis pump. He created it in such a way that it would stimulate more blood to the penis, so it would be thicker and would provide better erections. This has been the basis of every other penis pump that has been created since.

In the 20th century, some men started to rely on surgical advancements to help enlarge their penises, especially by cutting tendons that held the organ, so the penis drop would be longer than before. The problem with this though, was that it carried a lot of side effects, and men realized the surgeries were just not worth it.

The methods for penis enlargement have varied throughout the ages, but the reasons for seeking to enhance the size of the penis remain the same:

1. To appear "manlier" than others, thereby claiming the right to lead or be served.

2. To enhance sexuality by developing the ability to prolong sexual climax for as long as possible.

Many psychologists believe this extreme focus on the male anatomy is the primary cause for many of the wars waged by men throughout history. Some believe that, due to man's innate need to determine who has the bigger member, man has to be able to keep the upper hand.

In due time, after laws were created to prohibit the waging of unnecessary wars, men found a new way of besting each other—by showing who had the bigger penis.

Needless to say, males in the 21st Century are probably quite glad they no longer have to resort to any of these methods.

Chapter 3:

Best Practices in Jelqing

If you still have not figured it out, jelqing has a lot to do with stretching. Basically, you have to pull the penis away from the body, in such a way that you'll be able to hold it for a couple of seconds in an outstretched position. There may also be times when you could flex the muscles found in the base of the penis and then stretch by holding onto it to create tension.

The main purpose of jelqing is to lengthen the penis. This could happen with the help of the division of the cells in the penis, together with stressing penis tissues, in order for it to become longer. You can also make motions of pulling the penile ligaments out (even without really pulling them out, in a literal sense), so the inner parts of the penis will be exposed, and you'd feel much stronger as well.

Some Reminders

One thing to remember before engaging in the technique of jelqing is the importance of applying just the right amount of pressure to the strokes. Thinking along the same lines of milking a cow's udder, it is quite practical to note that applying too much pressure while stroking would hurt both the udder and the penis; this is the reason males are advised to apply only a moderate amount of pressure while stroking their penises in jelqing.

There are a few other things you have to keep in mind before performing any jelqing exercises. These are:

1. Make sure not to perform any jelqing exercises without warming up first.

2. Make sure to use lubricants—unless you're going to try dry jelqing - but do not use soap, as this literally dries up the penis. You'll see some of the best lubricants listed below. Aside from lubricants, you can also use rubber gloves, baby powder, a cloth, baby socks (with toe ends cut off) to aid in performing better jelqs.

3. It's always best to have a semi-erect penis before jelqing. However, you can also do it on a fully erect penis, as long as you have had enough practice. Jelqing on a semi-erect penis means you'd slowly be able to help your penis adjust to the exercise, and soon enough, it would have more threshold for pain and for any other kinds of exercises to which you might subject it to.

4. There might be the so-called "doughnut effect" or red spots on the penis during the first few jelqing attempts. The doughnut effect happens when the foreskin seems to be swollen or puffy for those who have not been circumcised, while it's the corona that gets this way when one is circumcised. Irritation may also occur not only while jelqing, but also while using penis pumps.

5. Take note that after a jelqing session, your penis would definitely look longer and thicker than normal. Of course, this might boost your ego, but you have to understand that, eventually, the length will subside, and your penis might shrivel. Don't worry, though, because this is normal, and it should not be a cause of alarm. It is similar to the "pump" you might experience after resistance training to your body, only to see the muscles return to their normal size later that day.

6. You have to remember that, sometimes, it might be hard to jelq when you're not aroused or while not ejaculating. In order to deter erection and experience the full benefits of jelqing, you should go ahead and squeeze your penis mildly.

Hand Positioning

The position of your hands is always an important part of any jelqing technique. The two most common hand positioning techniques are the following:

The OK Grip

This is the most common hand position, wherein you make an "OK" sign between your index finger and your thumb.

The Pincher Grip

On the other hand, you can also make use of the Pincher Grip. This way, you'd place your thumb on the bottom of your penis and your index finger on top of it. Now, you should see a line being formed with the help of your thumb and index finger, and your palm should face the top of your penile shaft. Make sure both index finger and thumb are pointing down.

The more extensive steps for jelqing involve:

Warm up the penis. This is an important step, so the penis can react in the right way, once the strokes are applied. This warm-up step can be performed by carefully slapping the penis back and forth. While lightly slapping, make sure the penis touches the walls of the inner thighs. This step is best performed right after a bath.

If slapping the penis feels uncomfortable, males can also perform this warm-up step by draping a small cloth over their penises. Males have to make sure the cloth is soft enough not to cause any abrasions, yet warm enough to drape over the penis and still warm it up.

Once the penis has warmed up a bit, it is time to move on to the next step. This involves grasping the penis, using the index finger and the thumb of one hand. As advised in the previous chapter, it is best to start from the base of the penis and move up towards the neck of the penis, stopping just below the corona. Lightly squeeze the penis with both fingers and make sure none of the other fingers get involved, at any point, during the exercise.

Slowly, slide the fingers along the shaft and release them upon reaching the corona or the head. However, make sure to get the fingers of the other hand ready at the base of the shaft before releasing the head. Once the head is released, start with the same stroke and squeeze in the same manner, using the other hand. This repetition can be referred to as one Jelq set.

Repeat the Jelq set about 15 times or more, depending on the level of endurance for beginners. It normally takes about 15-30 minutes for the 15 sets to be completed. This timeframe provides an advantage for males, who lead busy lives and cannot afford to spend hours with male enhancement therapy.

As soon as the jelqing session is complete, males can simply clean up and move on to their day's itineraries. Do the sets again on the following day after bath time. Beginners may find a little bit of difficulty mastering the strokes during the first few sessions. However, given time and enough repetition, they should be able to master it and endure the strokes, without ejaculating, until the sets are completed. One thing to remember is, jelqing works best when the penis is already semi-erect before the first stroke, hence the warm-up step emphasized earlier.

For beginners who wish to lubricate their penises, they may do so, since there haven't been any known adverse effects of lubrication while performing the jelqing technique. However, it is best to remember that soap is never a good lubricant, because it can contain harsh chemicals, which may cause too much friction, thereby making it painful for the penis to be stroked. The best lubricants to use are:

Cocoa butter

Petroleum jelly (Vaseline is a primary choice among many males.)

Baby oil

Oil-based lotions

It is also important to remember you shouldn't use lubricants that provide too much sexual stimulation, since this might result in premature ejaculation. This would shorten the jelqing session and make it harder for the individual to re-start the session.

Although lubrication may be preferred by some men, it is still possible to perform the jelqing technique on a dry penis. As a matter of fact, some men might prefer dry jelqing strokes, especially if they have overly sensitive penises that could ejaculate with just a few strokes.

Aside from lubrication, another thing to take into consideration when doing the jelqing method, is the stroke. Although the method emphasized in the steps above is the most popular way to jelq, men can also do the strokes using the pincher method.

This involves positioning the thumb and index finger, in such a way that it looks as if they could be picking something up off the floor with just two fingers. Contrary to the "O" method, where the two fingers are parallel to each other during the strokes from the base to the tip, in the pincher method, the index finger has to be placed on top of the penis.

In some cases, males may use both strokes, first using the pincher grasp to achieve a semi-erection and then moving on to the "O" grasp for the Jelq sets. Again, the grasp would depend, completely, on the male's comfort.

Beginners to the jelqing technique usually are advised to start at a slow pace in order to get their bodies to adjust to the exercise. They can perform the following suggested repetitions at the beginning:

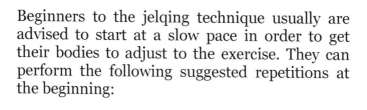

Two to three hundred repetitions per day in the first week.

Three to five hundred repetitions per day in the second week.

As they move on to the third week, they can start doing the standard five hundred repetitions per day and even more if they can endure it.

Apart from the given examples, here are other methods of jelqing one can try:

Basic Stretching + Jelqing Routine

First, you should start with three sets of the following stretches:

Behind The Cheeks to the Right: 30-seconds

Behind The Cheeks to the Left: 30-seconds

Behind The Cheeks to the Center: 30-seconds

Followed by:

Straight Down Rotary Stretches: 25-Cranks

Straight Down to the Center: 30-seconds

Straight Down to the Left: 30-seconds

Straight Down to the Right: 30-seconds

And then:

Straight Out Rotary Stretches: 25-Cranks

Straight Out to the Center: 30-seconds

Straight Out to the Left: 30-seconds

Straight Out to the Right: 30-seconds

Lastly, perform the following:

Straight Up to the Right: 30-seconds

Straight Up to the Left: 30-seconds

Straight Up to the Center: 30-seconds

Straight Up Rotary Stretches: 25-Cranks

Remember, you should spend at least approximately 35 minutes for these to be done properly. After stretching, put some baby powder on your hands and then put some on your penis as well. You can make an "Okay" symbol with your forefinger and thumb, wrap it around the flaccid penis head, and go ahead and pull.

Make sure not to wrench or jerk it outwards. At first, you may feel like it's sinking, but in a span of 2 to 3 weeks, you may notice that your penis has lengthened noticeably.

The Ultimate Stretcher

1. Firmly grasp the head of your phallus and make sure not to wrap your hands around it so tightly that it loses circulation.

2. Pull the phallus and stretch it outwards with just enough force to feel the base of your phallus and what's going on inside the shaft.

3. Let it stay in that position for at least 10 to 30 seconds and then take a rest for a while before repeating.

4. Make sure to incorporate this into your daily life for at least 5 to 10 minutes each day.

Moving Backwards

This is a Jelqing method that would allow you to work in a sort of reverse order. Here's how:

1. First, pull the skin of your penis backwards with the use of your thumbs.

2. Let the other fingers provide support under the shaft and hold the position for around 10 seconds.

3. Take a little break and repeat. Do this around 5 to 10 minutes each day.

Thumb Stretcher

1. Use one hand to grasp the tip of your penis.

2. Hold the tip of your penis firmly and, again, do not make it so uncomfortable that you suffocate it and make it lose blood flow.

3. Now, use your other hand to hold the base of the penis and start pulling away, until you feel it stretching, and hold the position for around 10 to 15 seconds before releasing.

4. Take a short break and repeat what you have just done. Do this 5 to 10 minutes each day.

Opposite Pull

1. Grasp the tip of your penis and make sure it is flaccid.

2. Now, put your left hand on the beginning of the phallic base and then pull the penis, using both hands, in opposite directions. It may sound painful, but with enough control, it shouldn't hurt. Do this for around 10 seconds at a time.

3. Feel the inside of your shaft stretching and then take a short break and repeat. Do this for 5 to 10 minutes each day.

Dry Jelqing

As mentioned earlier, there is also something called Dry Jelqing. Of course, it's the same as normal jelqing, except that it would help you gain better control (at least, this is what those who have tried it are saying) over not turning your jelqing sessions into masturbating sessions.

Generally, you still have to do what's asked of you for normal jelqing sessions. The only difference is, you shouldn't allow your fingers to slide over your skin while jelqing, and the place where your fingers make contact should be the only ones making contact throughout the whole session.

This way, you can prevent the skin on your shaft from being too erect, which would prevent you from doing a viable jelq. When that happens, you still need to wait for the erection to subside before you continue, and of course this might take a chunk of your time.

However, if you've already been jelqing a lot and are now considered an "experienced jelqer", you'd notice there would already be loosened skin on your penis, which will make it easy for you to perform dry jelqs, even if the penis is totally erect. This sounds crazy, but it would be good for you to "listen" to your penis and see whether the pain is still comforting or if it's too much. If it's too much, know that it's okay not to continue.

Dry Jelqing is also said to be a cleaner way of helping your penis improve its length. It can also be done in places where jelqing would seem to be risqué. In short, even just lying in bed, you could dry jelq, whereas you still have to find a more private place for non-dry jelqing. Dry jelqing doesn't include any noisy sounds whatsoever, which might make it seem like a more comforting technique than the other, depending on one's circumstances.

Other Stretching Techniques

Apart from what was given earlier, you can also make use of the following techniques:

Bundles

Basically, you have to make one full turn in either a clockwise or counterclockwise position. At first, it may seem hard, but sooner or later, you'd realize you can actually make more than one full turn. When the penis is twisted, stretching happens, because it becomes tighter. You can also use this, together with V-Stretches.

V-Stretches

V-Stretches can be performed by using one hand to grasp the penis and then using your finger or any finger-like device, such as a pole or a stick, to press against the penis, so it will stretch out into a rudimentary V-shape. You can then apply pressure on top, on the side, or on the bottom of the penis.

Fowfer

Lastly, you can try fowfer stretching. This means you'd pull the penis through your legs or between the buttocks and then sit on it. This can be painless and intense when performed properly.

Depending on the person, some methods may seem easy, and others challenging. You can try each of them out and see what you're comfortable with.

Chapter 4:

Advantages of Jelqing

Penis enlargement is probably one of the most sought-after, biggest advantages of jelqing. A large percentage of males prefer it over other male enhancement techniques, because it is an all-natural method, which does not require the use of artificial enhancers or the intake of any drug. Experts and supporters of the jelqing method have also cited a few more benefits including:

As a man, it's a good way of being in touch with your body. When you jelq, you get to realize you don't always have to drink pills or go through the lengths of having surgical procedures done, just so your penis would reach the length you want it to have. In short, you can feel like you're actually capable of a lot, and you can control your body in more ways than you think.

The texture of the penis improves. This can be attributed to the fact that the regular stroking involved in the jelqing method promotes the growth of a thicker and tougher shaft. Your penis may develop a more youthful look.

Another great thing is that jelqing gives you max power. It's said that jelqing is one of the healthiest things you can do for your penis, because aside from increasing length and girth, it also increases blood flow, which has a lot of amazing benefits, one of which is making sure you get better stamina, so you last longer while making love to your partner. This way, you can be sure you're really able to satisfy your partner, aside from being satisfied yourself, even without the use of lube.

The penis becomes more resilient and elastic. This means it becomes stronger, and therefore, less prone to premature ejaculation. This also means whatever penile growth was achieved through the jelqing exercise would remain as it is, even if the individual stopped doing the exercise for an extended period of time.

As an added bonus, sexual performance is also improved, since the penis is already used to prolonged erections without ejaculation. This restraint is also attributed to the improvement in blood flow that goes through the veins in the penis and scrotum.

Jelqing gives you more sexual control. Sexual control is important, because it allows you to be more in tune with your body; it makes you feel powerful, because you wouldn't seem like someone who's overly sexed, or one who doesn't even feel aroused at all costs. You'd be a person who knows what he's doing in bed—someone who can really please someone else. This way, your partner would also be more excited to make love and enjoy intimacy with you—and won't avoid you or feel like you're the only one who's going to benefit.

The improvement in blood flow to the male genital area also results in one more advantage for men: the increase in the production of testosterone. Every human being above puberty age is probably aware that testosterone is the hormone responsible for all the behaviors and characteristics that are unique to the male species. This means, an increase in testosterone level, through jelqing, can make a man healthier in his bodily functions, as well as his appearance.

One benefit of jelqing for enlarging the penis was confirmed in a study that involved one thousand Czech women. These women were asked questions that verified whether the size of the penis made a difference in the quality of their orgasms. A large percentage of these women provided affirmative answers to the correlation between a large penis and a great orgasm. These findings have probably provided sufficient reason for males of all ages to engage in exercises that could enlarge their penises.

Jelqing is also cost-effective. You don't need any expensive materials to do simple jelqing. You only need your hands. You wouldn't have to pay for things that might lead to complications, unless you really feel like you need extra help. It's one of the most normal ways to help yourself feel better, and that could really improve the state of your body.

However, males have to keep in mind, they are only able to take advantage of all the benefits that jelqing can provide if they have already been doing the exercise for some length of time. An enhanced penis that is resilient and has a high level of endurance cannot be achieved overnight. Patience and perseverance are always the most important keys to success.

A person looking for these results, would have to give his penis time to adjust to the exercises. One way to make the exercise easier to endure, without ejaculating prematurely, is for the male to do it at the time of day when he is least stressed, such as before going to bed at night or after taking an early morning shower.

Chapter 5:

Disadvantages of Jelqing

Just like any other form of therapy, jelqing also has its share of disadvantages. One of the known side effects of jelqing, which actually has caused a lot of worry for many men, is the occurrence of red spots on the penis.

Many believe these tiny red spots are caused by the tiny capillaries inside the penis that have burst, due to the application of too much pressure. This can happen if the penis has not yet adjusted to the jelqing exercises and/or is not warmed up properly. This is the reason beginners are advised to take it slowly and make sure their initial sets of strokes are firm, yet gentle, so they can avoid the red spots. This also happens when you make use of too many stretching exercises before jelqing itself.

Aside from that, you can also expect that ligament pops may occur in the consecutive weeks after doing your first jelq. To prevent this from happening, you should be more cautious with the way you stretch and jelq, which means you have to make use of the LOT Theory.

The LOT Theory is all about figuring out the right angle you should use for stretching. The theory says you need a higher angle to "tug back" when the ligaments are short, tight, or high. This is also what you should keep in mind when making use of Kegels. This needs to be done, so you access your "inner" penis more, and so you could get easy, no-fuss gains right away. It's like being able to stretch without feeling that extra itch, which will make the technique more comfortable and natural for you.

During the course of the entire jelqing therapy, there is still a possibility that the red spots may appear every now and then. This should not be a cause for worry, since the spots normally disappear a few minutes after the exercise is finished. However, should the spots start to turn blue or purple, instead of fading away, males are advised to check in with their doctors as soon as possible. This is another situation where the importance of warming up proves important.

Another known side effect of jelqing is referred to as the "doughnut effect", wherein seminal fluid gets clogged up at the tip of the penis while jelqing. This happens differently for circumcised and uncircumcised men. Because the structure of a circumcised male's penis involves a bare and bald glans, this is where the seminal fluid, sometimes, becomes clogged.

There may also be times, especially in the first few weeks after your first jelqs, you might experience soreness and penis insensitivity. This is because your penis is adjusting to the technique. You could prevent this from happening by making sure you easily pull back the skin from the glans, instead of pulling the skin that's around the penis itself.

On the other hand, an uncircumcised penis would still have a few centimeters of protruding foreskin that covers the glans, and this is where the fluid lodges. This can be quite painful, for both circumcised and uncircumcised males. However, just as the red spots disappear once the exercise is over, the clogged fluid of the doughnut effect will subside by itself within a few hours.

One of the reasons lubrication is recommended by specialists for males who wish to try jelqing, is so they can avoid the appearance of stretch marks. Dry jelqing, or jelqing without any form of lubrication, has been known to cause the appearance of stretch marks on the penis in some cases.

Besides, in some males, dry jelqing has the possibility of turning from merely uncomfortable to totally painful. This can happen when the lack of moisture between the penis and the fingers causes too much friction. And, as everyone knows, friction becomes hot, and too much heat on the genitals is painful. Unlike red spots and the doughnut effect, stretch marks have a tendency to stay visible on the penis, even after the exercise has finished.

Also, in the same way that side effects of other therapies and methods have cures and workarounds, jelqing does as well. Just as people who undergo bouts of hardcore muscle workouts treat their aching muscles with a nice warm bath after each workout, the penis can be treated the same way.

Some experts suggest wrapping the penis with a moist, warm towel after the exercise session to help with recovery. Once the towel has gone cold, the individual can rub some moisturizing fluid onto the penis, which would effectively ease any lingering discomfort.

Finally, there are very few cases where jelqing leads to impotency. Again, these are limited or isolated cases only, and usually it occurs when men use pills that are not FDA-approved, along with jelqing.

Remember: These things are not meant to scare you. They're only posted to help you understand what could happen while jelqing, so you know what you are getting into.

Chapter 6:

Male Enhancement in the Modern World

If you have done any research on the topic, you probably know the jelqing technique is not the only male enhancement therapy available in today's world of rapidly-evolving technology. There are many other options from which males can choose, although jelqing has retained its spot on the list of the most widely-used forms of male enhancement therapy in the world. However, it is still important to learn about other options if you are serious about male enhancement. Some other popular options include:

Pills

There are probably hundreds of male enhancement pills on the market today. However, the three most popular brands are LastimateX, VigRX Plus, and Prosolution pills. These three have ratings of above 78% effectiveness and are noted as the fastest in producing results for their clients. Surveys have also shown that these three brands have received the highest ratings in customer satisfaction, with LastimateX leading the race, with at least 98.4% customer satisfaction.

Almost all of the male enhancement pills are made of a few common ingredients, such as primary compounds, like Diamana and Epidium. Another ingredient common to most pills is ginseng, an herb that has been used in male enhancement since the time of the pharaohs.

These pills could enhance the way your penis lengthens, especially while you're jelqing. However, it's best to make sure you don't take too much of them, so you don't suffer side effects, such as nausea, digestive pain, and even mood swings!

Some of the highest rated male enlargement pills in the market include:

Vimax

Vimax is made with ginseng and gingko biloba, and it is meant to provide better performance and increased sex drive. More importantly, it provides firmer erections.

Pro-solution

Aside from providing firmer erections, Pro-solution also increases one's sex drive and can provide more intense orgasms.

VigRX Plus

While it doesn't provide permanent growth in the size of one's penis, one can expect VigRX to provide temporary firmer erections, increased orgasms, and better sex drive.

ExtenZe

ExtenZe also does the same as the examples above, but what's notable is, those who have used it say their partners have actually felt that the thrusts they make are much more powerful than before.

Make sure, though, you do not use brands that are not approved by the FDA, nor those you can get from the deep web. Remember, if you decide to go this route, it's still best to use those that have been approved and that you know have been tried by many. Make sure to consult with your doctor before consuming any male enhancement pills.

Physical Exercises

Aside from jelqing, there are a few other types of exercises devised to increase the girth of the penis via physical repetition. One such exercise is stretching. The first step of the exercise is to ensure the penis is flaccid and then lift it gently into a horizontal position. The penis is then stretched gently and held in this position for about 30 seconds, then brought back to the original flaccid position.

This exercise only has to be performed in 5 repetitions, unlike the hundreds of strokes involved in the jelqing exercise. However, both the jelqing and stretching exercises share the common goal of maximizing blood flow to the penis in order to enlarge it.

Surgery

The different types of surgeries today for male enhancement are not as painful or unsafe as the penile surgeries of the past. One such modern form of surgery is known as phalloplasty.

Originally, phalloplasty referred to the process of either constructing a male penis in a transgender female or the reconstruction of the penis of a natural-born male, who suffered from a genital disease. The term eventually evolved to include penis enhancement. The enhancement of the penis is done in different ways, such as the implantation of an inflatable penile device that would increase the size of the penis. The implant is often referred to as a prosthesis, and it is used not only for enhancing penile size, but also for other medical purposes.

Another type of male enhancement surgery involves cutting off the ligaments that connect the internal portion of the penis to the underside of the pubic bone. Once the ligament has been detached, the portion of the penis that normally remains hidden is able to move up and out, thereby elongating it. This type of surgery is not as popular as other male enhancement surgeries, since it carries more potential risks, like cutting the incorrect ligament, which could cause further problems with erection and sexual performance.

Surgery to enhance penile size can be expensive, and not every man can afford it, which is the reason male enhancement surgery is often the last resort for men, who are looking for ways to enhance their size. On the other hand, male enhancement pills may have adverse side effects, and even if this is a remote possibility, most males would rather not take the risk. This is probably the reason natural methods for male enhancement, such as jelqing, have remained popular, in spite of all the modern technology at everyone's disposal.

Men, whose primary objections to jelqing stem from the fatigue their fingers *might* experience due to all that stroking, are advised to try using jelqing devices. These devices take the place of the fingers in stroking the semi-erect penis. The devices would not only serve to ease the pain from tired hands, but ensure that just the right amount of pressure is being applied to the penis with each stroke.

Men do not have to worry about the potential embarrassment and disapproving stares while buying the device, since they can simply order it from an online store. These stores ensure 100% anonymity to their customers, even during the delivery of the product.

Jelqing Devices

Jelqing devices also promise faster results than doing the basic two-finger method. However, this claim is *not* backed by any data, as of yet, so its truthfulness can only be judged by those who have already tried jelqing with and without a device.

And with jelqing devices, price should not be a problem, since most of these devices are sold at less than a hundred dollars each. They also come with money-back guarantees to assure customers of the product's effectiveness. Some examples are:

Hydro Pumps

These are said to be some of the best penis pumps, because they distribute water evenly around the surface of the phallus, making the experience more comfortable for you.

Vacuum Pumps

Vacuum Pumps provide the feel of jelqing without having to do it with your own hands. It is also said these might treat erectile dysfunction in the long run. Though these claims have not been validated at this time.

Extenders

True to its name, extenders help you stretch your penis the way you want it to be stretched. Tension is gradual, but is said to provide amazing benefits in the long run.

Remember: Do not consume any of these items until after consulting with your doctor!

Conclusion

I worked hard on creating the best introductory guide for jelqing that I could. The next step is to put everything you have learned from this book into good use. It is never too late to work on self-improvement, whether through physical fitness, expanding your knowledge base, or from genital exercises in the bathroom. Before trying methods that could burn a hole through your pocket, it is best to try the all-natural methods first.

If you decide to take the path of jelqing, do not hesitate to look around for an expert in jelqing who would be able to answer questions that might arise. This way, you can cover all possible grounds and pave a clear path towards a sexier, more virile version of yourself.

If you feel like you learned something from this book, please take the time to share your thoughts with me by sending me a message or posting a review to Amazon. It would be greatly appreciated!

Thank you and good luck on your journey!

Made in the USA
Middletown, DE
29 November 2018